CONTENTS

Foreword

This guide has been prepared for:

- designers
- developers
- occupiers
- employers
- fire authorities
- building control bodies.

It explains the steps involved in approving the fire safety aspects of building work, and the interaction between Building Regulations and other statutory fire safety requirements in England and Wales.

If the procedures described in this guide are followed, developers and designers will be able to receive all the fire safety information relevant to Building Regulations and other statutory approvals, in time to avoid abortive work. Owners and occupiers will also benefit, as this process will generate the basis of their fire safety management procedures and risk assessment. ***There should be no need for extra building work to be done at the end of a building project, before the building can be occupied for its intended purpose.***

It is intended that all building control bodies and fire authorities should use the consultation procedures described in the guide as a model for arrangements they make, so that procedures will be similar throughout England and Wales.

The Department of the Environment, the Welsh Office, the Home Office and the Deregulation Unit of the Department of Trade and Industry issued the first edition of this guide jointly in 1992. This document is a revision of the original guide to take account of recent changes to fire precautions legislation and associated procedures.

Although this guide has no legal force the Department of the Environment, Transport and the Regions, the National Assembly for Wales and the Home Office would expect relevant bodies and authorities to follow its recommendations. The recommended procedures will be reviewed periodically and amended in the light of any changes in legislation.

Terminology

The following terms are used in the guide:

applicant – the person responsible for obtaining approval. In practice this is often the agent of the owner or developer, such as the architect, and it is used in this guide to include anyone acting for the applicant.

building control body – a term used to include both local authority building control and approved inspectors.

> **local authority** – the local authority empowered to carry out the building control function and to enforce Building Regulations in their area by virtue of Section 91(2) of the Building Act 1984.
>
> **approved inspector** – a corporate body or individual approved under Section 49 of the Building Act 1984 to carry out certain building control functions.

consultation – exchange of information and comment between organisations, such as building control bodies and fire authorities, which may include correspondence and face-to-face meetings.

fire authority – the authority discharging the functions of the fire authority under the Fire Services Act 1947 in the area in which the premises are, or will be, situated. In most cases the enforcement responsibilities of fire authorities is delegated to the fire brigade.

relevant use – means a use as a workplace of a kind to which Part II of the Fire Precautions (Workplace) Regulations 1997 applies **or** a use designated under Section 1 of the Fire Precautions Act 1971(see appendix A).

Approved Inspectors Regulations – The Building (Approved Inspectors etc.) Regulations 2000 (S.I. 2000/2532, ISBN 0-11-099898-7, £6.30, The Stationery Office Ltd).

Building Regulations – The Building Regulations 2000 (S.I. 2000/2531, ISBN 0-11-099897-9, £4.50, The Stationery Office Ltd).

Workplace Fire Regulations – The Fire Precautions (Workplace) Regulations 1997 (S.I. 1997/1840, ISBN 0-11-064738-6, £3.20, The Stationary Office Ltd) as amended by the Fire Precautions (Workplace)(Amendment) Regulations 1999 (SI 1999/1877, ISBN 0-11-082882-8, £2.50, The Stationery Office Ltd).

SECTION 1 – INTRODUCTION

WHAT LEGISLATION APPLIES

1.1 This guide is mainly concerned with the Building Regulations which apply only in England and Wales. There is separate legislation covering building in Scotland and Northern Ireland. Appendices A, F, G and H give information on other legislation that applies to fire safety, including buildings in use. The main procedural points are covered in the step by step guide in section 2.

Building Regulations

1.2 The Building Regulations apply to building work as defined in regulation 3 of the Regulations. However, section 4 of the Building Act 1984 provides exemptions for some educational establishments (see appendix G) and some buildings belonging to statutory undertakers, the United Kingdom Atomic Energy Authority and the Civil Aviation Authority. Schedule 2 of the Building Regulations exempts various types of building according to their use or size. Buildings belonging to, or that are occupied by, the Crown authorities are also exempt.

1.3 Fire safety requirements are given in Part B of Schedule 1 to the Regulations. These cover means of escape, fire alarms, fire spread, and access and facilities for the fire service. Guidance on some ways of meeting the requirements is given in Approved Document B (Fire Safety).

Fire Precautions Act 1971

1.4 Premises put to the following uses:

- hotels or boarding houses
- factories
- offices
- shops
- railway premises

are designated, subject to certain exceptions, under Section 1 of the Fire Precautions Act 1971(see appendix A), and will require a fire certificate (see appendix B) unless specifically exempted by the fire authority.

Workplace Fire Regulations

1.5 These Regulations apply to premises in which persons are employed (see appendix A). The fire authority is responsible for the enforcement of this legislation (with the exception of ships under construction or repair and any workplace, other than a building on the surface at a mine, to which the Fire Certificates (Special Premises) Regulations 1976 apply - see appendix G). By virtue of these Regulations the employer is required to undertake and continually review a fire risk assessment and implement appropriate precautions to ensure that the employees within the premises are not placed at risk from fire.

Parallel Application

1.6 Most (not all) buildings subject to the Fire Precautions Act are also subject to the Workplace Fire Regulations. Some may also be subject to other legislation imposing fire safety requirements. Most building work, and certain changes of use, involving buildings that are subject to fire precautions legislation will also be subject to the Building Regulations.

1.7 The frequent parallel application of the fire safety requirements of the Building Regulations and other legislation relating to fire precautions emphasises the need for consultation between the applicant and the administrating bodies involved at the earliest opportunity.

1.8 If the consultation procedures described in this guide are followed there should be no need for extra building work to be done after the end of a building project before a building can be occupied for its intended purpose. This aim can only be achieved where all parties take a methodical joint approach to fire safety, whilst discharging their respective statutory duties and responsibilities.

THE DIVISION OF RESPONSIBILITY

1.9 There are two main bodies with whom designers, developers and occupiers of buildings may have dealings concerning fire safety: the building control body and the fire authority.

1.10 Building control bodies are responsible for checking on compliance with the requirements of the Building Regulations. The Regulations are concerned with building work and with material changes of use (which may give rise to requirements for building work) and the requirements for fire safety will apply to most buildings.

1.11 The fire authority is the enforcing authority for matters relating to fire safety under the Fire Precautions Act 1971, the Workplace Fire Regulations and either directly or indirectly other legislative controls which concern the safety of people in relation to the operation and use of certain buildings once occupied.

1.12 Building work that complies with the Building Regulations' requirements for fire safety will normally be satisfactory when the building is occupied. However, there may be risks associated with the specific operations of the occupier that would not be covered by Building Regulations. There is obvious practical value in taking these other fire safety matters into account in the design and construction phases, although the legislation only becomes applicable on occupation of the building.

1.13 During the design and construction phase of a project the building control body will check on compliance with the requirements of the Building Regulations. In order to facilitate the consultation process they should take a **co-ordinating role** with fire authorities and, where appropriate, with other regulatory bodies. Any recommendations and advice given should be channelled through the building control body to the applicant.

1.14 Once a building is occupied and in use the fire authority take on the co-ordinating role as they have the enforcement role for general fire safety matters under the Fire Precautions Act 1971 and the Workplace Fire Regulations. Licensing and registration authorities will also consult the fire authority about fire safety matters concerning premises that need to be licensed or registered. Fire authorities are not themselves generally the licensing or registration authority. Note that licensing/registration conditions are also likely to refer to matters other than fire safety. Appendix G summarises legislation under which licensing or registration may involve consultation on fire safety.

BUILDING REGULATIONS – COMPLIANCE IN GENERAL

1.15 It is the responsibility of anyone carrying out building work to comply with the relevant requirements of the Building Regulations. The Approved Documents associated with the Regulations give practical guidance on how to comply. The design and construction of building work is subject to checks by a building control body. Applicants can decide whether to apply to the local authority for building control or to appoint an approved inspector. The two systems of building control are detailed in the Manual to the Building Regulations.

Local Authority Building Control

1.16 Where the applicant chooses to use the local authority there will generally be a choice of two routes: the 'full plans' procedure, or the 'building notice' option. A 'full plans' application involves seeking formal approval of plans. Approval or rejection must be given within a five week time limit (or two months if the applicant agrees), and there is a facility to request a determination of disagreements from the Secretary of State (see appendix D). When a full plans application is made the applicant can ask the local authority for a completion certificate when the work is complete. A building notice, however, is simply given to the local authority at least two clear days before work begins. There is no associated right of appeal (other than to a magistrates' court) against any requirement the local authority may make, should they decide that there is a contravention when they inspect the work. **If the building is to be put to a relevant use the building notice procedure cannot be used (see appendix A).**

Approved Inspector Building Control

1.17 If an approved inspector is engaged the person intending to carry out the work and the inspector must jointly give to the local authority an initial notice. The local authority has five days in which to accept or reject the notice. If the local authority have neither accepted or rejected the notice by the end of that period, the notice is deemed to be accepted.

1.18 Once the notice has been accepted the approved inspector is responsible for supervising the work. If the applicant wishes to have detailed plans of the work certified as complying with the Building Regulations, and the approved inspector is satisfied with the plans, a plans certificate will be issued to the applicant and the local authority by the approved inspector. There is a facility to request a determination of disagreements from the Secretary of State (see appendix D). When the work is complete the approved inspector must give the local authority a final certificate.

RISK ASSESSMENT AND THE WORKPLACE FIRE REGULATIONS

1.19 By virtue of the Workplace Fire Regulations employers are required to carry out a fire risk assessment of their workplace. This must consider all employees and all other people including the public and persons with special needs who may be affected by a fire and may be present in the premises. (Detailed guidance on compliance with these regulations is contained in "Fire Safety – An Employers Guide" published by HSE Books – ISBN 0-11-341229-0).

1.20 Although these Regulations are applicable to a workplace whilst in operation it would be useful for the designers of a building to carry out at least a preliminary risk assessment as part of the design process. The level of detail of this assessment will vary depending on the amount of information available to the designer and whether or not the eventual occupier of the building is known at the design stage.

1.21 If a preliminary risk assessment is produced it can be used as part of the Building Regulations submission and can assist the fire authority in providing advice at an early stage as to what, if any, additional provisions may be necessary when the building is first occupied. It will also assist employers when developing the full risk assessment for the purposes of the Workplace Fire Regulations, and it will act as a record of the rationale behind the fire safety design of the finished building.

1.22 It is fundamental to the principle of the Workplace Fire Regulations that fire precautions will usually need to change and evolve with the building throughout its life. Risk assessments should be kept constantly under review to make sure that precautions are adequate and effective.

OCCUPIED BUILDINGS

1.23 Responsibility for ensuring that a building, once occupied, is provided with appropriate fire safety arrangements rests with the occupiers, owners and employers, according to the circumstances. It is therefore essential that, when occupying new buildings or existing buildings that have had building works completed, or when contemplating alterations or extensions to existing buildings, occupiers (and owners/employers where appropriate) ensure that the fire safety arrangements in place are adequate and meet the requirements of any relevant enforcing authority (see appendices A, B and G).

1.24 Ensuring that these responsibilities are met is achieved through a number of different systems, enforced by one or more statutory authorities, which may involve the certification, licensing or registration of the premises concerned. It is possible that a building will be subject to more than one set of regulations, or other legislation, relating to fire safety and that more than one enforcing authority may be involved.

1.25 However, with the exception of most dwellings and a small number of other cases, the fire authority will have a primary enforcement responsibility, or a responsibility for advising the enforcing authority concerned. The fire authority's responsibilities are contained in a number of different pieces of legislation. These range from the direct enforcement responsibilities contained in legislation such as the Workplace Fire Regulations and the Fire Precautions Act 1971, to that of statutory consultee/advisor to the enforcing authority contained in such legislation as the Registered Homes Act 1984 and the Licensing Act 1964.

1.26 It is therefore recommended that occupiers (and owners or employers where appropriate) who are unsure what legislation applies to their building, or who require advice on the fire safety arrangements made, contact the fire authority for the area concerned. On being contacted the fire authority will advise on the legislation that, in their opinion, applies to the premises concerned and the appropriate enforcing authorities. In those cases where the fire authority has a direct or advisory role they will advise on the procedures that should be followed and the fire safety arrangements that should be in place or maintained.

1.27 When considering the fire safety provisions for the building in use the fire authority will have regard to the risk assessments that were undertaken as part of the design and approvals processes and by the employer under the Workplace Fire Regulations, and the building control body's completion or final certificate. The consultation between the designers/applicant, the building control body and the fire authority, and the recording of risk assessments and other relevant information, throughout the design and approval process should ensure that the subsequent need for a variation in the fire safety arrangements provided is avoided, or kept to a minimum.

1.28 Additional fire safety arrangements should only become necessary because of factors identified on occupation of the building that were not made known, or were unknown, at the time of the building regulation approval and consultation process. It is, therefore, essential that the fire authority is fully consulted on all relevant matters and that their advice on the fire safety arrangements for the building when in use is incorporated, if appropriate, into the final building design.

1.29 If in the effort to achieve an acceptable fire safety package the views of the fire authority and the building control body differ, and cannot be resolved, the building control body should ensure that the applicant is fully aware of the possibility that the fire authority may require additional works on occupation of the building(see 2.17.1).

CONSULTATION

1.30 When a building is likely to be used for a purpose that is subject to the fire safety requirements of other legislation, consultation between the relevant bodies is an essential part of the building control procedure. Whilst there are clear distinctions of jurisdiction between building work and an occupied building it is often impossible when considering the overall level of safety in a building to separate physical fire safety measures and the way in which the building will be managed when occupied. This guide describes the statutory consultations that fire authorities and building control bodies are obliged to carry out. It also indicates where discussions may be needed to consider fire safety proposals at an earlier stage than those required for statutory consultation so as to keep all parties informed and avoid delays. These discussions may be between the applicant and the building control body only, or they may also involve the fire authority.

1.31 **Where a building control body is formally engaged in checking the compliance of building work with the Building Regulations and the building in question is to be put to a relevant use there will be statutory requirements to consult the fire authority at certain stages of the process.**

1.32 These consultations are necessary so that the fire authority is aware of the erection or alteration of buildings in which they may have a direct or indirect interest or responsibility. They also provide an opportunity to make the applicant aware of action that may have to be taken on non-building regulation matters in order to satisfy any other fire safety legislation once the building is occupied.

1.33 Fire authorities and building control bodies will often hold useful information relating to the design of existing buildings. The consultation process provides an opportunity to share this information with the other parties.

1.34 There may be occasions where more than one building control body is dealing with different building work within the same building. In these circumstances it would be advisable for them to consult with each other, as necessary, to avoid any conflict that might otherwise arise.

1.35 The nature and timing of consultations are described in detail in section 2.

1.36 Consultation with all relevant parties can also help to achieve consensus where alternative technical solutions may be available, such as during the specification of fire detection and fire alarm systems. As an example, a fire risk assessment, agreed by the relevant parties, may assist in determining the fitness for purpose of alternative means of providing a fire alarm e.g. a shouted warning or a simple electrical alarm or a fault monitored electrical alarm. Effective decisions made at the specification stage can save considerable rectification costs if corrective action proves necessary after completion of the building work.

1.37 There are also circumstances where a fire authority, in the course of its own statutory activities, would consult building control bodies. Such circumstances include consultation under the Fire Precautions Act 1971 (Section 17) prior to serving a "steps notice" (see B4 appendix B), or under the Workplace Fire Regulations (regulation 13(5)) prior to the serving of an "enforcement notice" which would oblige a person to make an alteration to a building.

CONSULTATION – OTHER LEGISLATION

Local Acts

1.38 There are provisions in other primary legislation, and in some places in Local Acts (see appendix H) which require consultation to take place between local authorities and fire authorities. This guidance should also be used in those instances. Under regulation 13(6) of the Approved Inspectors Regulations an approved inspector must consult the fire authority if a Local Act would have required the local authority to do so, had they been undertaking the building control function. The applicant should check with the building control body to see whether there are any Local Acts in force. These Acts are usually enforced by the local authority only.

Premises Subject To Licensing Or Registration

1.39 Many premises are controlled through a licensing procedure in which fire safety matters are an important concern. Where plans are submitted to the building control body for new buildings, or alterations or extensions, which may be subject to licensing or registration, the building control body should consult all interested parties (see appendix G).

Buildings that may require registration or licensing could include:

- Residential/Nursing Care Homes
- Hospitals
- Boarding Schools
- Public Houses
- Places of Entertainment
- Night-clubs

- Nurseries
- Sports Grounds
- Places of Marriage
- Houses in Multiple Occupation
- Petroleum Filling Stations.

Special Premises

1.40 The Health and Safety Executive has enforcement responsibilities under the Fire Certificates (Special Premises) Regulations 1976 for general fire precautions in special premises (premises such as nuclear installations, explosives factories and magazines or where hazardous substances are stored or used in quantities above specified limits).

SECTION 2 – THE APPROVALS PROCESS STEP BY STEP

2.1 This section describes the consultation procedures that should be followed at different stages of the building control process.

2.2 During these stages the building control body is the co-ordinating authority, as explained above, and their consultation with the fire authority is crucial. It is very important that all parties should respond within agreed timescales, and that they should play an active part in maintaining good communications.

2.3 The consultation process should be conducted to ensure that both the building control body and the fire authority fulfil their roles in an efficient and cost-effective way. Where local arrangements exist which already fulfil the aims of this guide it may not be necessary to adopt all the procedures suggested here.

2.4 The plans approval process is likely to be delayed if designs are presented without prior consideration of fire safety. Where local authorities undertake building control there are statutory time limits on the period for consideration of plans and so delays may lead to rejection of plans. Throughout this guide it is assumed that applicants are conversant with technical aspects of fire safety and its regulation, or that they employ professional advisors who have this knowledge. **Whilst early and effective consultation is strongly recommended, the advice given by the building control body and the fire authority does not extend to design consultancy.**

The left-hand column describes actions of direct concern to the designer or applicant. The right hand column provides commentary or describes actions which are of less direct concern to the applicant. Although the process is presented in a chronological sequence this may vary in practice, and some of the stages will be not be relevant in all cases.

PRELIMINARY DESIGN STAGE ADVICE AND CONSULTATION

2.5 At the early stages of a project the designer may wish to seek advice about the fire safety aspects of the scheme. Involving building control bodies and fire authorities at an early stage can often result in reduced costs for the applicant. However the advice given by the building control body and the fire authority does not extend to design consultancy. The designer's first approach should preferably be to the building control body.

2.5.1 Whatever body is contacted first should determine, as far as it is able, which other authorities or bodies are likely to have an interest and inform the applicant of the necessary application or consultation procedures. Therefore, at the earliest opportunity the applicant should be informed of the likely interest of the other organisations and the necessary application or consultation procedures.

2.5.2 In responding to an independent approach from a designer or occupier, for goodwill advice under the Fire Services Act 1947 (see appendix F), the fire authority should point out that in respect of the Building Regulations, or other legislation for which they are not directly responsible, they can offer only informal opinions. They should refer the applicant to a building control body, and where appropriate other enforcing authorities, for guidance on what may be required to meet the legislation for which those bodies or authorities are responsible.

2.5.3 Any advice the fire authority do give should be in writing and should clearly indicate which matters in their opinion:

- *will have to be complied with under the Fire Precautions Act 1971 or the Workplace Fire Regulations when the building is occupied.*
- *will have to be complied with to meet other fire safety legislation.*
- *are only advisory and not enforceable under legislation. A copy of the advice should be sent to the local authority and the approved inspector where it is known that a relevant initial notice is in force.*

2.6 If the building or part of the building is to be put to a relevant use the fire authority will have powers that may influence the design of the building (see appendix A).

2.7 In such cases, if there are aspects of the design which follow an unusual or complex approach to fire safety, it may be desirable for the designer to request a joint meeting with the fire authority and the building control body.

2.8 At this meeting the building control body should take the co-ordinating role. A record of the main points of the meeting should be circulated to all parties, which clearly distinguishes between:

- building regulation requirements
- fire certification requirements
- requirements of any other fire legislation
- advice which is not enforceable under legislation.

STATUTORY CONSULTATION

2.9 Where a building is to be put to a relevant use, building control bodies are required to consult with the fire authority at certain stages within the process as prescribed by legislation.

2.9.1 Whilst the legislation and associated procedures for local authorities and approved inspectors may be different the purpose of consultation is the same, that is to seek the comments of the fire authority regarding the fire precautions that will be necessary to meet the legislation (other than the Building Regulations) that will apply to the building once it is in use. This should allow them to reach ***mutually compatible*** *views on whether plans and building work are satisfactory from the standpoints of the Building Regulations and of fire precautions legislation.*

Note

In addition to giving such comments the fire authority may wish to offer ***observations*** *to the building control body in relation to the Building Regulations. These should be clearly and separately identified.*

2.9.2 Consultation also provides the opportunity for the fire authority and the building control body to make recommendations regarding the provision of fire precautions that it believes would be advisable but that are not required by legislation.

2.10 Even if there is no formal requirement to consult, consultation may still be desirable in many cases such as:

- where the building control body proposes to accept measures which are unusual or complex and/or
- where the size or location of the development might have implications for the disposition of fire service resources.

2.11 To facilitate consultation the building control body will need two extra copies of those drawings that demonstrate compliance with Part B. In some cases the two extra copies will be needed for statutory consultation purposes. In cases where no statutory consultation is involved the local authority or approved inspector may still choose to consult the fire authority. In either case the two extra copies allow the fire authority to keep a set of drawings while returning the other to the building control body, marked up with any comments.

2.11.1 Under regulation 14 of the Building Regulations if plans are being submitted to a Local Authority for building work to which Part B of Schedule 1 to the Building Regulations is applicable, unless the building concerned is a house or flat, two extra copies of those plans which demonstrate compliance with Part B are required to be provided by the applicant for consultation purposes.

2.11.2 Under regulation 13 of the Approved Inspectors Regulations, an approved inspector, when consulting in relation to an initial notice, must provide the fire authority with a set of plans which demonstrate compliance with Part B. When consulting in relation to a plans certificate an approved inspector must provide the fire authority with a copy of all the plans to which the certificate will relate. Approved inspectors will need to make arrangements to ensure that sufficient copies are available.

2.12 Where plans are deposited with a local authority for approval and the intended use of the building is a relevant use the local authority is required to consult with the fire authority under section 16 of the Fire Precautions Act before passing or conditionally passing the plans.

2.12.1 Section 16 of the Building Act allows local authorities 5 weeks, or 2 months if agreed in writing, to pass or reject plans. Fire authorities must, therefore, respond in good time(see 2.16).

2.13 Where approved inspectors have been appointed and the intended use of the building is a relevant use they are required to consult with the fire authority before or as soon as is practicable after issuing an initial notice and before giving a plans certificate or final certificate to the local authority.

2.13.1 Approved inspectors may not give a plans certificate or final certificate until 15 working days have elapsed from the date on which they consulted with the fire authority, unless the authority have replied before the end of the 15 days (regulation 13 of the Approved Inspectors Regulations).

2.14 If the building control body is not satisfied that the proposals comply with Building Regulations, they may send the applicant a list of amendments or additional information considered necessary. They will not normally consult the fire authority until they are reasonably satisfied that compliance with Building Regulations is shown. If the deficiencies are minor they may decide to proceed with consultation, attaching a copy of their list of the necessary amendments (clearly identified as such) with the information they send the fire authority.

2.14.1 If the proposals need substantial amendment or the information provided is inadequate then the process may be delayed. It is therefore very important that the application shows that full account has been taken of Part B's requirements, if such delays are to be avoided.

2.14.2 In some cases a local authority may be prepared to consider relaxing or dispensing with a requirement under regulation 11 of the Building Regulations in which case additional consultation may be required (see appendix E).

2.15 Once the building control body is reasonably satisfied that the proposals comply with the Building Regulations they will send to the fire authority two copies of drawings which relate to compliance with Part B, i.e. which identify the intended physical fire precautions, together with any supporting documentation they consider useful. That might include correspondence they have had with the applicant. It may also be helpful at this point for the building control body to advise the fire authority of any unusual aspect of the design.

2.15.1 Building control bodies should time their consultations and provide such information to the fire authority as necessary to satisfy the purpose of the consultation (see 2.9.1).

2.15.2 If the building control body is aware of any information that may assist the fire authority or if they have given goodwill advice on additional precautions that are advisable but not enforceable under legislation (see appendix F) then the building control body should include this information with the consultation documents.

2.16 The fire authority should make its observations to the building control body in writing and within agreed timescales (usually within 15 working days) so that the building control body can meet its own obligations.

2.17 The fire authority's comments must clearly distinguish between matters:

- which will have to be complied with under the Fire Precautions Act 1971 or considered under the Workplace Fire Regulations when the building is occupied.
- which will have to be complied with to meet other fire safety legislation **other than Building Regulations**.
- which are only advisory and not enforceable under legislation (see appendix F).

Note

In addition to giving such comments the fire authority may wish to offer **observations** to the building control body in relation to the Building Regulations. These should be clearly and separately identified.

2.17.1 If in the effort to achieve an acceptable fire safety package the views of the fire authority and the building control body differ, the building control body should seek to ***resolve the matter quickly and simply****. Failing this the fire authority should indicate its views in a formal written case to the building control body, who should, if necessary, convene a meeting to* ***achieve consensus****.*

2.17.2 In the ***exceptional*** *event that the fire authority and building control body cannot agree, the DETR, in consultation with the Home Office, is prepared to give informal advice where it is approached jointly by both parties. Any approach to the Department should be in writing and should include a statement of the case from each party taking into account all relevant legislation.*

2.18 The building control body must have regard to the fire authority's comments before reaching its decision on plans or in the case of an approved inspector before deciding on a final or plans certificate. In giving their decision building control bodies should include a copy of any comments from the fire authority (see 2.22).

2.18.1 Although fire authorities should make their observations in writing the fire authority's comments in whatever format will be of interest to the applicant and should be passed on by the building control body.

APPROVAL OF PLANS

2.19 Having plans approved by either the local authority or certified by an approved inspector can give protection from enforcement action where the approved plans have been followed.

2.19.1 A local authority cannot give a notice under Section 36 of the Building Act 1984 requiring alterations if building work conforms to plans that have been approved by the local authority. In cases where an initial notice ceases to be in force without a final certificate having been given, the local authority may not give a notice under Section 36(1) in relation to work that conforms to plans that are the subject of a plans certificate issued by the approved inspector.

2.20 When, after any necessary consultations with the fire authority, the building control body is satisfied that the plans comply with the requirements of Building Regulations they may issue a notice or certificate approving the plans.

2.20.1 Where a full plans application has been submitted to a local authority they must issue a decision notice within the statutory time limit. They may reject the plans, approve them or issue a conditional approval.

2.20.2 In cases where an approved inspector is undertaking building control, if an applicant requests a plans certificate, the approved inspector must give one if the approved inspector is satisfied that the plans comply with the Building Regulations (see 2.13.1).

2.21 If a statutory consultation was involved, the building control body should also send a copy of the notice or certificate (or some other mutually acceptable written notification) to the fire authority.

2.22 The building control body should supply the applicant with a copy of the comments and advice generated by the fire authority described in paragraph 2.17.

2.22.1 In the ***exceptional*** *event that the fire authority propose to require physical changes in plans or work that are beyond the requirements of the Building Regulations the building control body should make the applicant aware.*

AMENDED PLANS

2.23 On many projects the design of the building will change during both the design and construction stages. Plans may be amended a number of times before and after the statutory consultation stage is reached. If the applicant submits amended plans, following a statutory consultation and before or after the plans are approved, because of changes relating to fire safety made to the scheme, the building control body should, as a matter of good practice, consult the fire authority, even if no question arises of formally approving plans.

2.23.1 The fire authority should be sent two copies of amended plans, one to put comments on and one to keep. Any comments from the fire authority should be in writing, and should follow the format described in paragraph 2.17, and should be relayed to the applicant by the building control body.

2.24 Where an approved inspector has been appointed some proposed changes in a building project may require the giving of an amendment notice to the local authority, altering the description of the work given in the original initial notice. The approved inspector will then need to formally consult with the fire authority again, under regulation 13 of the Approved Inspectors Regulations, if work introduced by the amendment notice concerns a building to be put to a relevant use and is subject to Part B requirements.

2.24.1 The procedures for statutory consultation detailed above should be followed. It may be possible to streamline this procedure if the fire authority has been kept up to date during the design process.

CONSTRUCTION

2.25 As work proceeds the building control body may make inspections as appropriate.

2.26 The purpose of these inspections is to assess the works as they progress with regard to compliance with the Building Regulations, which relate to fire safety and other requirements of the Building Regulations for the **finished building only.**

2.27 Building Regulations do not address the risk of fire during the construction work, which is covered by the Construction (Health, Safety and Welfare) Regulations 1996.

2.27 HSE have issued the following fire safety guidance on these regulations: Construction information Sheet No 51 Construction Fire Safety; and HSG 168 Fire safety in construction work (ISBN 0-7176-1332-1).

2.28 When construction work is being carried out on an existing building which, apart from the construction site part of the building, remains occupied, the fire authority are responsible for the enforcement of the Construction (Health, Safety and Welfare) Regulations in respect of fire. Where the building is unoccupied, the Health and Safety Executive are responsible for enforcement.

2.28.1 The fire authority enforces parts of the Construction (Health, Safety and Welfare) Regulations relating to fire safety matters in existing buildings undergoing construction works. As a result fire authority officers may, from time to time, inspect premises undergoing works of construction to ensure that the fire safety provisions and measures are and remain adequate for the risk presented.

COMPLETION

2.29 Whilst this section logically falls before occupation it is not unusual for a building to be either fully or partly occupied before it is formally recognised as complete. Paragraphs 2.36 to 2.41 give guidance on the occupation of buildings.

2.30 The applicant should notify the building control body that the building work is complete.

2.30.1 Where a local authority is supervising the work regulation 15(4) of the Building Regulations requires that the local authority should be notified of the completion of the building work not more than 5 working days after completion. Approved Inspectors should advise the applicant what notification arrangements should be followed.

2.31 At this stage the applicant should produce an updated risk assessment and drawings to record the fire safety provisions included in the building design.

2.31.1 Whilst it is not a statutory requirement to produce "as built" information, it will often be in the client's best interests to do so. These benefits are illustrated in paragraphs 1.19 to 1.22.

2.32 If the building control body is satisfied that the work complies with the requirements of the Building Regulations, they should issue either, in the case of a local authority, a completion certificate or, in the case of an approved inspector, after consultation with the fire authority a final certificate.

2.32.1 There is no obligation on a local authority to issue a completion certificate unless one has been formally requested, or unless the building will be put to a relevant use.

2.32.2 Approved inspectors must give the local authority a final certificate following completion of the work, if satisfied that the work complies with the applicable requirements of the Building Regulations. Approved inspectors are subject to a statutory time limit for the issue of a final certificate starting from occupation (see 2.37.2).

2.33 Approved inspectors must consult with the fire authority before issuing a final certificate (regulation 13 of the Approved Inspectors Regulations). Local authorities, as a matter of good practice, may also consult at this stage. Such consultations need not necessarily involve additional plans.

2.34 The building control body should send a copy of the completion/final certificate to the fire authority including where available a copy of the risk assessment and/or "as built" record drawings.

2.34.1 The fire authority should have regard to the completion/final certificate when considering an application for a fire certificate or reviewing the fire safety provisions of an occupied building under the Workplace Fire Regulations or other fire safety legislation.

2.35 If, having been notified of completion, the building control body cannot issue a completion certificate or final certificate, they should explain, in writing, why this is so.

2.35.1 A copy of this explanation should be sent to the fire authority where fire issues are involved. If necessary the fire authority and the building control body should meet to discuss any inadequacies that are thought to exist in the means of escape or structural fire precautions and consider what options are available for advice or enforcement.

OCCUPATION

2.36 Prior to occupying the building the owner/employer must ensure that where relevant the following statutory obligations are met.

Building Regulations

2.37 If it is proposed to occupy a building where a Building Regulations completion or final certificate has not been issued (see 2.29 to 2.35) by the building control body then the building control body must be notified of the timing and extent of the occupation in advance.

2.37.1 Where a local authority is supervising work involving the erection of a building, regulation 15 of the Building Regulations requires the applicant to give the local authority at least 5 working days prior to occupation. In accordance with Section 92 of the Building Act 1984 any such notice should be in writing.

2.37.2 Where an approved inspector is engaged and an initial notice relates to the erection, extension or material alteration of a building that is to be put to a relevant use and the building (or relevant part)is occupied and no final certificate is given, the initial notice will cease to have effect after a grace period of 4 weeks (8 weeks for non relevant uses) and building control will usually revert to the local authority, unless the local authority agree to extend the period.

Workplace Fire Regulations

2.38 If the building is a workplace (see appendix A) the employer must have completed the fire risk assessment and the provisions required to address the identified risks must be in place.

2.38.1 There is no period of grace for employers to produce their risk assessment. The documentation and any necessary safety measures must be in place on the first day that the building is occupied.

2.38.2 There may be additional risks associated with the logistics of moving an organisation into a new building that should be addressed by the employer.

2.39 If the designer has produced a risk assessment as part of the design and approvals process this may form the basis of the employer's assessment.

Fire Precautions Act

2.40 If the building is being put to a use where a fire certificate is required (see appendix A) the application for the certificate must be with the fire authority before occupation.

2.40.1 Whilst there is a statutory requirement for the application for a fire certificate to be made there is no need for a fire certificate to be in place before a building is occupied. There is however, a statutory duty which falls upon the owner/occupier to maintain certain fire safety measures in the period between a fire certificate being applied for and being issued. Similarly the Workplace Fire Regulations may also apply, in which case the employer must carry out a fire risk assessment.

Licensing and Local Acts

2.41 If the proposed building is subject to a licence or a local Act there may be conditions applicable to occupation of the building (see appendices G and H).

2.41.1 If in doubt the fire authority is a useful source of advice as to the applicability of fire safety, licensing and local legislation.

APPENDIX A – RELEVANT USES

BUILDING REGULATIONS

A.1 For the purposes of the Building Regulations and the Approved Inspectors Regulations a "relevant use" is defined as meaning use as a workplace to which Part II of the Workplace Fire Regulations applies or a use designated under section 1 of the Fire Precautions Act 1971.

FIRE PRECAUTIONS ACT 1971

A.2 The use of certain types of premises has been designated by the Secretary of State as requiring a fire certificate under the Fire Precautions Act. There are two designating orders in force in Great Britain.

A.3 The first designation order requires a fire certificate when premises are used as a hotel or boarding house which will provide sleeping accommodation for more than six people (whether staff or guests), or if they provide any sleeping accommodation for staff or guests elsewhere than on the ground or first floors of the premises.

A.4 The second designation order requires that a fire certificate must be applied for when more than 20 people are at work at any one time in a factory, office, shop or railway premises, or more than 10 are at work at any one time elsewhere than on the ground floor.

A.5 In buildings in multiple occupation containing two or more similar premises, a certificate must be sought when the aggregate number of people exceeds the totals in either A3 or A4 above.

A.6 Fire certificates are also required for factory premises (these may include certain warehouses) where explosive or highly flammable materials are stored or used in or under the building, regardless of the number of people at work, unless the fire authority has determined otherwise.

A.7 The fire authority may exempt premises from the certification requirement if they consider them to be of low risk.

WORKPLACE FIRE REGULATIONS

A.8 Part II of the Fire Precautions (Workplace) Regulations 1997 applies (subject to certain exceptions) to any premises or part of a premises, **not being domestic premises**, used for the purposes of an employer's undertaking and which are made available to an employee of the employer as a place of work.

CONSULTATION

A.9 Regulation 17(3)(a) of The Workplace Fire Regulations modifies section 16 of the Fire Precautions Act 1971 to the effect that references there to "a designated use" shall include reference to use as a workplace.

A.10 The amendment of section 16 of the Fire Precautions Act 1971 does not have the effect of extending the requirement to apply for a fire certificate, which remains unchanged.

A.11 **Under regulation 12(3) of the Building Regulations, the "building notice" procedure identified in regulation 12(2)(a) cannot be used in respect of any proposed building work relating to a building to be put to a relevant use. Therefore in the case of work to a building which is to be put to any such use, full plans must be deposited, if the local authority system is selected. The local authority must then consult the fire authority in accordance with Section 16 of the Fire Precautions Act 1971 (as amended).**

A.12 The consultation requirements in regulation 13 of the Approved Inspectors Regulations are likewise tied to the concept of relevant use.

APPENDIX B – FIRE CERTIFICATE

B.1 The intending occupant or owner of a building must apply to the fire authority for a fire certificate before any premises are put to a use for which a fire certificate is required (see appendix A). The fire authority is required to carry out an inspection.

B.2 If, following the inspection, the fire authority is satisfied that the means of escape from fire and related fire precautions in the premises concerned are such as may reasonably be required in the circumstances of the case, it must issue a fire certificate.

B.3 The consultation process should ensure that the building can be certificated without further work. But if the fire authority is not satisfied that the fire precautions meet the requirements of the Fire Precautions Act 1971 it will, subject to the restrictions of the statutory bar (see appendix C) serve a notice on the applicant for the fire certificate requiring steps to be taken to rectify matters, and specifying the time for completion of those steps.

B.4 Under Section 17 of the Fire Precautions Act 1971 the fire authority will consult the local authority before serving a notice if alteration work is required. Generally this consultation entails sending a copy of the proposed notice to the local authority. Where the fire authority is aware that an initial notice is in force with respect to the building, or that part of the building, to which the notice relates it may be appropriate to additionally consult the approved inspector.

B.5 The fire certificate, or in the case of premises forming part of a building in multiple occupation, a copy of the certificate must be kept in the premises to which it relates. It is an offence not to keep the fire certificate in the premises.

B.6 The fire certificate will specify:

a the use(s) of the premises covered by the fire certificate.

b the means of escape in case of fire.

c the means for ensuring that the means of escape can be safely and effectively used at all material times (this would cover such matters as measures to restrict the spread of fire, smoke and fumes, escape lighting and directional signs).

d the type, number and location of the means for fighting fire (whether in the premises or affecting the means of escape) with which the relevant building is provided.

e the type, number and location of the fire alarms.

f in the case of factories, particulars of any explosive or highly flammable materials which may be stored or used in the premises.

B.7 The certificate may include a plan showing any of the above.

B.8 The fire authority may also decide to incorporate any of the following requirements in the fire certificate:

a that the means of escape are properly maintained and kept free from obstruction.

b that everything covered in (c), (d) and (e) in paragraph B6 is properly maintained.

c that all employees are given appropriate training in what to do in case of fire, and that records are kept of that training.

d that the number of people who may be in the premises at any one time do not exceed a specified limit.

e other precautions to be observed in relation to the risk from fire to persons in the premises.

B.9 The Occupier (or owner in the case of multi-occupied buildings) of premises for which a fire certificate is in force, or where an exemption from the requirement to hold a fire certificate has been granted by the fire authority, has a duty to inform the fire authority of any proposed change of use and of any structural alterations or extensions.

B.10 If any of these changes are 'building work', as defined by the Building Regulations, an application should be made to a building control body and the procedures for consultation set out in section 2 should be followed.

APPENDIX C – THE STATUTORY BAR

C.1 The statutory bar applies to new or refurbished buildings, where plans have been approved under Building Regulations or an approved inspector's initial notice has been accepted and where the fire authority deem that the building requires a fire certificate under the Fire Precautions Act.

C.2 Plans should have been sent to the fire authority via the building control body. Having been consulted by the building control body the fire authority under the Fire Precautions Act 1971 cannot then require further fire safety measures for means of escape in the building on occupation, i.e. the statutory bar is applied.

C.3 However, if the fire authority is satisfied that the fire precautions are inadequate by reason of matters, relating to means of escape in case of fire, that were not subject to a Building Regulations approval or were not known at the time of approval, then the fire authority is not barred from making requirements.

C.4 The statutory bar does not apply to any requirements of the Workplace Fire Regulations. This reflects the need for flexibility in fire precaution assessments to evolve within the building throughout its life cycle. Fire risk assessment should be continually reviewed to ensure that the employees within the premises are not placed at risk from fire.

C.5 In practical terms in all but a few premises the statutory bar has ceased to be of effect. This makes it all the more important that consultation between building control bodies and fire authorities should result in a consensus on the fire safety aspects of building work at the earliest possible stage during the design and execution of a building project.

APPENDIX D – DETERMINATION

D.1 If the applicant and a building control body disagree about whether plans of proposed work are in conformity with Building Regulations, the applicant can seek a determination of the question from the Secretary of State under either Section 16(10)(local authorities) or Section 50(2)(approved inspectors) of the Building Act 1984

D.2 An application for a determination should be sent to the Department of the Environment, Transport and the Regions (or in Wales, to the National Assembly for Wales) who will charge a fee of half the relevant plans charge subject to a minimum of £50 and maximum of £500. The applicant should explain why they consider the proposal does comply, accompanying the statement of case with relevant drawings and a copy of any rejection notice.

D.3 There is no similar provision for determination by the Secretary of State under the Fire Precautions Act or the Workplace Fire Regulations.

APPENDIX E – DISPENSATION – STATUTORY CONSULTATION

E.1 The local authority is obliged by Section 15 of the Building Act 1984 to consult the fire authority before relaxing or dispensing with a requirement of Building Regulations that relates to:

- structural fire precautions
- the provision of means of escape from buildings in case of fire
- the provision of means for securing that such means of escape can be safely and effectively used at all material times.

E.2 Local authorities should also consult with the fire authority where they intend to relax or dispense with requirement B5 (access and facilities for the fire service).

E.3 The consultation should follow the form described in section 2 with the fire authority responding in writing.

APPENDIX F – FIRE SERVICES ACT, 1947

F.1 Section 1(1)(f) of this Act has particular relevance to this guide. It requires every fire authority to secure efficient arrangements for giving, when requested, advice in respect of buildings in the area of the fire authority as to:

- fire prevention
- restricting the spread of fire
- means of escape in case of fire.

F.2 During the consultation process a fire authority may offer goodwill advice to applicants that is not enforceable under legislation. It is important that this advice is clearly differentiated from advice relating to statutory requirements and that the benefits of adopting the advice is clearly explained.

APPENDIX G – OTHER RELATED LEGISLATION

Premises where other legislation requires notice to or consultation with the fire authority:

ANIMAL ESTABLISHMENTS

- Pet Animals Act, 1951
- Animal Boarding Establishments Act, 1963
- Riding Establishments Act, 1964 (as amended)
- Breeding of Dogs Act, 1973
- Zoo Licensing Act, 1981.

These premises are controlled by a licensing system operated by local authorities and, in any decision to grant a licence, the licensing authority is obliged to have regard to the protection of animals in case of fire or in an emergency.

CHILDREN'S HOMES

- Children Act, 1989.

Provision is made in Schedules 4, 5, 6, 8 and 9 for the making of Regulations covering fire safety in community homes, voluntary homes, registered children's homes, privately fostered children's accommodation and premises providing child minding and day care for young children. Regulation 14 of The Children's Homes Regulations 1991 (S.I. 1991/1506) makes various requirements of responsible authorities in relation to fire precautions.

ELDERLY AND DISABLED PERSONS HOMES

- Registered Homes Act 1984 (as amended by the Registered Homes (Amendment) Act 1991).

Regulations made under this Act – The Residential Care Homes Regulations 1984 (S.I. 1984/1345) and The Nursing Homes and Mental Nursing Homes Regulations, 1984 (S.I. 1984/1578)(as amended) – require persons registered to take adequate fire precautions in the home and to consult the fire authority for this purpose.

HOUSES IN MULTIPLE OCCUPATION

- Housing Act 1985 (as amended by the Local Government and Housing Act, 1989) sections 352, 365 and 368.

If a local authority believe that a house in multiple occupation is not provided with such means of escape and other fire precautions as the authority think necessary, they may serve notice specifying the work necessary to provide a proper means of escape and requiring work to be executed within a specified period. Before serving any notice etc. relating to means of escape the local authority must consult the fire authority.

LICENSED PREMISES

- Cinemas Act, 1985.

An applicant for the grant or renewal of a cinema licence must give written notice of the application to the fire authority. The licensing authority may grant, renew or transfer a licence having regard to any observations made to them by the fire authority.

- Gaming Act, 1968.

Applications for grant or renewal of a licence are made to the licensing justices and the fire authority has a right to be heard on any application made.

- Licensing Act, 1964 (as amended by the Licensing Act, 1988).

An application for grant or renewal of a justice's licence must give written notice of the application to the fire authority. The fire authority on the basis of fire risk can make objections to the grant of licence. In the case of licensed clubs, the fire authority has the right to make an objection to applications on the ground that 'the premises are not suitable and convenient for the purpose in view of their character and condition and of the size and nature of the club'.

- Local Government (Miscellaneous Provisions) Act, 1982.

An applicant for the grant or renewal of an entertainments license (for premises used for music, dancing and similar entertainments) has to notify the fire authority and, when considering the application, the licensing authority must have regard to any observations of the fire authority. This Act does not apply in London.

- Theatres Act, 1968.

The licensing authority may grant a licence on such terms and conditions as may be specified and, although conditions vary, they may relate to fire safety matters.

PLACES OF MARRIAGE

- Marriage Act, 1994.
- Marriages (Approved Premises) Regulations, 1995.

Section 1 of this Act amended the Marriage Act 1949 to permit civil marriages to take place on premises approved for the purpose by local authorities. Schedule 1 to these Regulations lists the requirements for the grant of approval, which includes fire precautions as may be reasonably required by the local authority having consulted with the fire authority.

PLACES OF SPORT

- Safety of Sports Grounds Act, 1975 (as amended by the Fire Safety and Safety of Places of Sport Act, 1987).

In London and the metropolitan counties, the local authority must send a copy of the application for a safety certificate to the fire authority and must consult them about the terms and conditions to be included in the certificate before it is issued. (Elsewhere, the copy must be sent to the building authority).

SCHOOLS

- Education Act, 1944 (as amended).

Some publicly maintained schools are exempt from the Building Regulations by virtue of section 4 of the Building Act, 1984 (as amended) where the proposed works are approved by the Secretary of State for Education. It is the intention of the DfEE to revoke regulations which currently permit approval of projects by the Secretary of State for Education and Employment and thus render maintained schools no longer exempt from the Building Regulations. Independent schools are not exempted from the operation of the Building Regulations. Also, under the Education (Particulars of Independent Schools) Regulations, 1982 (S.I. 1982/1730) the particulars of advice given by the fire authority on fire safety matters have to be declared in applications for registration of the school and in the annual return to the relevant Registrar.

SPECIAL PREMISES

- Fire Certificates (Special Premises) Regulations, 1976.

The Health and Safety Executive (HSE) has enforcement responsibilities under these Regulations for general fire precautions in special premises (premises such as nuclear installations, explosives factories and magazines or where hazardous substances are stored or used in quantities above specified limits). There is no statutory duty for building control bodies to consult with HSE but it may be beneficial in some circumstances for an informal consultation to take place

APPENDIX H – LOCAL ACTS

H.1 There are many Local Acts in operation in various areas in England and Wales and some of these include provisions requiring consultation with the fire authority. The local authority will advise where these Acts apply. Some of the provisions contained in these Acts may have been subsumed into Part B of the Building Regulations.

H.2 Some Acts where fire-related provisions occur are listed below:

Berkshire Act 1986

Bournemouth Borough Council Act 1985

Cheshire County Council Act 1980

Clwyd County Council Act 1985

County of Avon Act 1982

County of Cleveland Act 1987

County of South Glamorgan Act 1976

County of Kent Act 1981

Cumbria Act 1982

Derbyshire Act 1981

Dyfed Act 1987

East Sussex Act 1981

Essex Act 1987

Greater Manchester Act 1981

Hampshire Act 1983

Hereford City Council Act 1985

Humberside Act 1982

Isle of Wight Act 1980

Leicestershire Act 1985

Merseyside Act 1980

Poole Borough Council Act 1986

South Yorkshire Act 1980

Staffordshire Act 1983

Surrey Act 1985

West Glamorgan Act 1987

West Midlands County Council Act 1980

West Yorkshire Act 1980

Worcester City Council Act 1985

H.3 In addition to these Acts, there are fire-related provisions in the London Building Acts (Amendment) Act 1939, which applies in Inner London.

H.4 Typical fire-related provisions allow local authorities to impose conditions requiring access for the fire brigade, fire safety precautions in multi-storey car parks, fire precautions in tall buildings and large buildings used for trade or storage, and means of escape.